Penguin Nature Guides

Trees and Shrubs
of the Mediterranean

Helge Vedel

Illustrated by Roald Als and
Anette Rasmussen
Translated from the Danish by Aubrey Rush
Edited and adapted by Hugh Synge

Penguin Books

Penguin Books Ltd, Harmondsworth,
Middlesex, England
Penguin Books, 625 Madison Avenue,
New York, New York 10022, U.S.A.
Penguin Books Australia Ltd, Ringwood,
Victoria, Australia
Penguin Books Canada Ltd, 2801 John Street,
Markham, Ontario, Canada L3R 1B4
Penguin Books (N.Z.) Ltd, 182–190 Wairau Road,
Auckland 10, New Zealand

Træer og buske omkring Middelhavet first published by
Gyldendal 1977
This translation published 1978

Copyright © Gyldendalske Boghandel, Nordisk Forlag, A/S,
Copenhagen, 1977
Copyright © in the English edition: Penguin Books Ltd, 1978
All rights reserved

Printed in Portugal by Gris Impressores
Filmset in Monophoto Times by Northumberland Press Ltd,
Gateshead, Tyne and Wear

Contents

Foreword

The trees and shrubs in this book are those generally seen in the countryside and in the parks, gardens and open spaces of towns, attracting attention by their distinctive appearance – by the beauty of their flowers or the lushness of their fruit. The species described are typical of the countries surrounding the Mediterranean Sea; species regarded as familiar in our own latitudes are included only in a few isolated cases.

The book is intended as a guide for travellers interested in the flora of this part of the world. As such it is not exhaustive – the size of the book makes it impossible to include every species. (Those who wish to delve more deeply should consult the list of books on p. 125; these deal more extensively with the plant life of the Mediterranean.) It must be emphasized that this book deals only with trees and shrubs; large and showy plants like the banana and the agave are not included. The species are arranged in their botanical families.

The author and the artists acknowledge the help they have received in compiling this book, particularly from the Botanic Garden of Copenhagen University.

Agave

Introduction

Natural Environment and Climate

Evergreen forest covered most of the land of the Mediterranean region in prehistoric times. It is difficult to believe this today when the varied landscape consists of orchards, cultivated fields, grassy plains, mountains with scattered stands of pine, expanses of low brushwood and barren, stony desert. The landscape that we see today results from the activities of all the great civilizations which, over thousands of years, have inhabited the countries of the Mediterranean. Forests were cut down for timber and the scrub cut for firewood and for animal fodder; large areas were cleared for cultivation. These factors, combined with accidental and intentional fires, and with pastoral grazing, annihilated the old forest cover.

Sheep, and particularly goats, which can be adept at climbing trees, destroyed the young growth. In many places the plant cover vanished completely so that the bare fertile top soil was washed or blown away. In such areas the land has become barren and completely unproductive to man, a stark contrast to the plant-rich forests that originally covered the region.

Only a few scattered remnants of these forests exist today, for most of the forest has given way to areas of drought-resistant scrub known variously as maquis, garigue and steppe (see section on plant communities, p. 13).

Although the Mediterranean region is large enough for the climate to vary from one part to another, the main characteristic of a Mediterranean climate is simple: hot dry summers and mild moist winters. In such conditions, the Olive, the Holm Oak, the Kermes

Olive tree

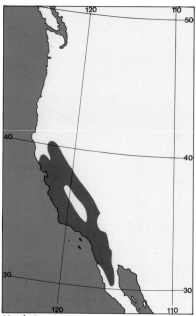

North America

Europe, N. Africa, Near East

South America

South Africa

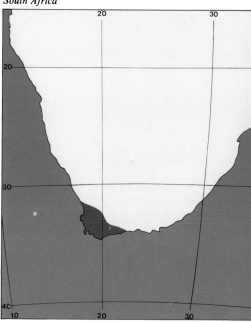

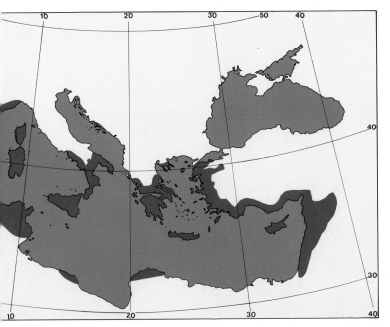

Regions with Mediterranean climate

Australia

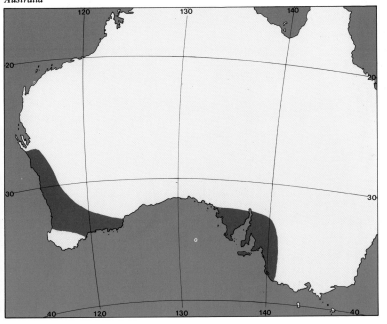

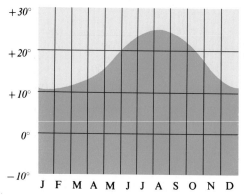

Graph showing average monthly temperature (°C) in Palermo (Sicily) from January to December.

Oak and the Aleppo Pine flourish and occur widely. Similar climatic regions are found in the western part of North America (mainly in California), parts of Chile in South America, the Cape region of South Africa and the south and south-west of Australia. In all these regions, although the plants differ from one area to another, the vegetation has tended to develop in a similar way to survive hot dry summers. (Pages 8 and 9 show where those regions with Mediterranean climate are to be found.)

The chart shows the monthly temperatures in Palermo, giving an average yearly temperature of around 17°C, considerably higher than that of Britain. Vegetation, however, is affected not only by the average temperature, but also by extremes. Frost, although rare in Mediterranean countries, is generally disastrous for plant growth, especially for those plants cultivated in the region.

The rainfall chart shows the total, month by month, in Palermo. Very little rain falls in the summer months and this, combined with the hot sun, day after day, is an effective impediment to plant growth unless, as in gardens, one can provide irrigation. Many of the crops of the Mediterranean,

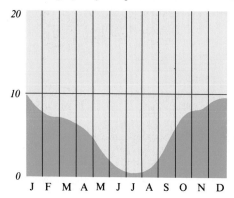

Graph showing average monthly rainfall (cm) in Palermo. There is least rainfall in the hottest months.

such as citrus fruits, have their growing season in the mild winter; the fruits ripen in the spring sun and are picked before the summer drought. In southern Greece, for example, the first oranges come about Christmas-time, and the crop is picked during the following months until late spring.

Plant growth depends on climatic conditions. From December to February, Almond, Peach and Apricot trees are in blossom along with plants growing from bulbs, corms and tubers. In March the rising temperature increases the number of flowering plants. From April to mid-May they are at their peak and include most of the species in this book. Especially conspicuous are the bright yellow brooms and the colourful shrubs of *Cistus*. But then follows the dry season continuing through July, August and September. Towards the end of September the autumn rains begin, stimulating new activity – from then until December the Strawberry Tree, the Carob and autumn bulbs are in flower.

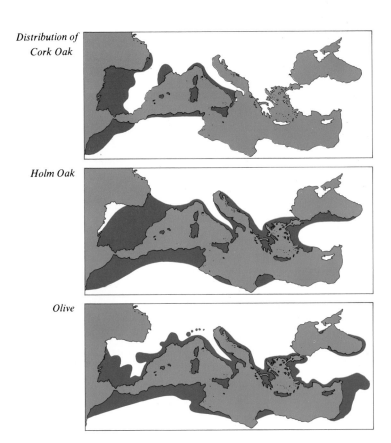

Distribution of Cork Oak

Holm Oak

Olive

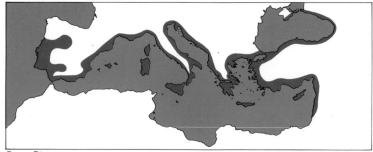

Stone Pine

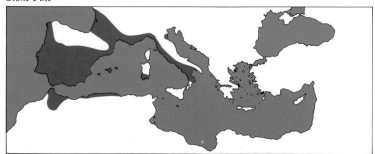

Maritime Pine

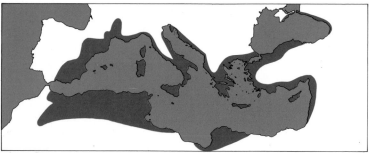

Aleppo Pine

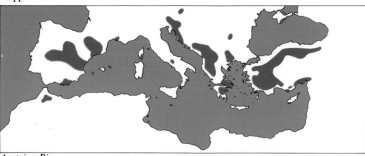

Austrian Pine

12

Plant Communities

The remains of the old forest cover, which can still be found here and there, can be divided into three types.

Evergreen Forest

The most widespread is the evergreen forest, dominated by the Holm Oak (*Quercus ilex*), the Cork Oak (*Quercus suber*) in the western regions, and changing to a more scrub-like growth with Kermes Oak (*Quercus coccifera*) in eastern countries. Fully grown Holm Oak forest is 10–15 m high, but seldom reaches this size. It is dense with undergrowth of Buckthorn and *Viburnum*, Strawberry Tree and *Clematis*.

Forests of Cork Oak are more open, with a rich undergrowth of shrubs and other plants.

The most important of the forest conifers are the Stone Pine, Maritime Pine, Aleppo Pine and varieties of Austrian Pine.

Pines tend to grow more on the northern, less arid side of the Mediterranean; Aleppo Pine, Maritime Pine and Stone Pine are all common in coastal districts. Austrian Pine and Grecian Fir, in contrast, are more trees of the mountains where they form extensive forests. However, the Maritime Pine does extend inland, and grows on many of the mountains of the Iberian Peninsula. Aleppo Pine has the widest distribution of all the Mediterranean pines, growing both north and south of the Mediterranean Sea, presumably since it can tolerate very dry conditions.

Maquis

Maquis is the first stage in the degeneration of evergreen forest – a process begun centuries ago. A

low scrub community consisting of small trees and bushes 2–4 m high, tolerant of seasonal drought, the maquis derives its name from the Corsican word for a species of the Sun Rose (*Cistus*) – a striking component of the maquis community. Others are Tree Heather, Strawberry Tree, Lentisc and various brooms. Many, such as Rosemary, Rue, Mint, Thyme and Sage, are strongly scented, a factor which helps to protect them from grazing animals. Dominant plants are often stunted trees and bushes of Juniper or Kermes Oak. The maquis is intensively used – as a source of fuel, and leaf-fodder for animals, dye for clothing, tan for leather dressing, resin and rubber, briar root and a wide variety of materials for household use. Most of the wild plants illustrated in this book flourish in the maquis and are useful to man in many ways.

Garigue

As the grazing, felling and cutting intensifies, the maquis becomes reduced to a community of low, rounded shrubs, one half to one metre high, in an expanse of bare earth, stones and rocky outcrops.

Garigue is found, in large areas, chiefly in the driest and hottest regions. Due to the intense over-grazing, many of the shrubs become stunted and only the most spiny tend to survive. The small annuals and bulbous plants usually grow underneath the bushes, where they are partly protected from grazing.

There is not much difference between maquis and garigue – rather they are stages of degeneration – and so many of the same plants occur in both communities, especially the Kermes Oak and the strongly scented small shrubs.

Garigue in early spring is a mass of flowers from bulbous and tuberous plants such as Iris, Crocus and Tulips; but the flowering is soon over. Then the garigue becomes parched and dusty.

Steppe

If the cycle intensifies even further, the garigue can be reduced to a dry steppe in which all the shrubs have disappeared, and only a few annuals and tuberous plants remain. Of these the most characteristic are the colourful Asphodel and Sea Squill.

Plants and Climate

Cold winters in temperate regions are unfavourable to plant growth: in the Mediterranean region, however, it is the hot, dry summers that are the limiting factor. The plants of the maquis, garigue and steppe have adapted in several different ways to summer conditions, principally to protect themselves against water shortage and evaporation. For example, the Holm Oak has tough, leathery leaves which do not collapse and wither in a drought. Some species secrete wax to trap moisture within the leaves. Others have densely hairy and felted leaves. In drought-resistant plants the pores in the leaves (stomata), through which evaporation takes place, are often small and sunk beneath the leaf surface (epidermis). Many plants of the heather family have involuted leaves which protect the stomata.

The smaller the leaf's surface, the less evaporation takes place; this is seen at its extreme in plants such as Spanish Broom and some of the acacias.

To get the benefit of any possible water in the soil many plants have deep and wide-ranging root systems, far more extensive than their surface growth. Succulent species such as *Euphorbia canariensis* and Prickly Pear are particularly well adapted, and store water in their roots and thickened stems. Plants with underground stores of food in bulbs and tubers are typical of many dry regions. They rapidly produce leaves and flowers and then seed in the early spring or late autumn, when there is moisture, and in the dry season the surface growth withers away completely. Other plants, such as annuals, survive the hot summer as seed.

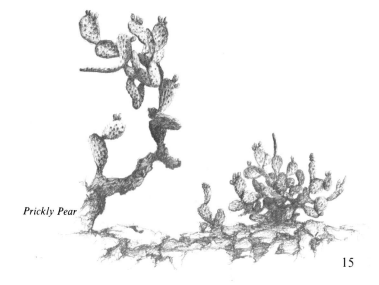

Prickly Pear

Individual Plants Described and Illustrated

The countries of the Mediterranean are an ideal place to see a great variety of plants. The flora is more extensive than in the countries of northern Europe. Spring, with its abundance of flowers and its pleasant climate, is an ideal time to visit the region.

The plates which follow cover the majority, but not all, of the more common trees and shrubs you are likely to see. Beside the pictures are short notes on each plant. For some the notes show how the species can be distinguished from another plant closely related to it; in these cases the differences will work for common Mediterranean species, but may fall down in other regions where their relatives are different. Technical terms have been avoided – perhaps the only one whose meaning will not be immediately clear will be the word 'pinnate': a pinnate leaf is one divided into leaflets arranged like the hairs of a feather, i.e. in two opposite rows down an unbranched axis. The illustrations of *Cassia* on p. 60 show this type of leaf.

One of the pleasures of the Mediterranean flora is the large number of species introduced by man from other lands, either as crops or as ornamentals. Many have long been established in the Mediterranean and have since become naturalized, that is, they have seeded themselves into the wild where they now persist without aid from man. Trees like *Eucalyptus* are now such a dominant and well-known feature of the landscape that it is easy to forget they are not part of the native flora.

Another pleasure is that most of the plants have long associations with man. As will be seen from the pages which follow, a great many of the trees are of interest in some particular way, or are the source of some useful and valuable product. Many of these applications go back for centuries, and, indeed, are perhaps less appreciated today than they were in the past. In an age when so much of the world's flora is in danger of extinction, and when our main crops become more and more precariously based on a smaller and smaller range of plants, an appreciation of the rich and varied flora of the Mediterranean, and how it has been fundamental to man's life and civilizations in the region, can only enhance and strengthen the arguments for widespread conservation of plants. This is true not only in the Mediterranean but also in those regions where the flora, although perhaps equally diverse, has only been partially explored, and whose full benefits are as yet unknown.

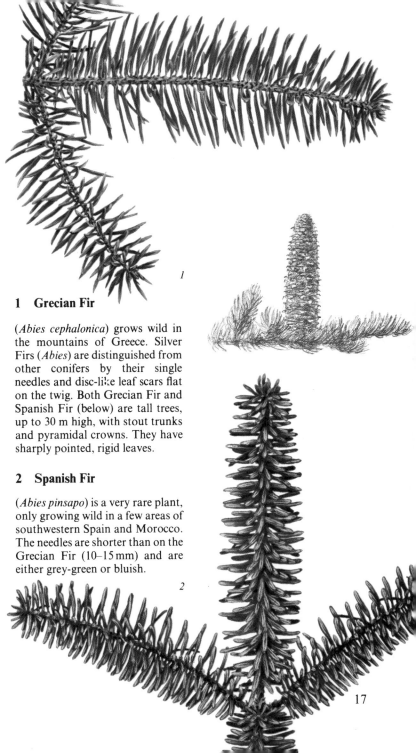

1   Grecian Fir

(*Abies cephalonica*) grows wild in the mountains of Greece. Silver Firs (*Abies*) are distinguished from other conifers by their single needles and disc-like leaf scars flat on the twig. Both Grecian Fir and Spanish Fir (below) are tall trees, up to 30 m high, with stout trunks and pyramidal crowns. They have sharply pointed, rigid leaves.

2   Spanish Fir

(*Abies pinsapo*) is a very rare plant, only growing wild in a few areas of southwestern Spain and Morocco. The needles are shorter than on the Grecian Fir (10–15 mm) and are either grey-green or bluish.

17

3 Oriental Spruce

(*Picea orientalis*) is native of Asia Minor, Armenia and the Caucasus. It has very short (6–10 mm), shiny, dark green needles, nearly square in cross-section, which, on falling, leave small pegs protruding from the twig (in contrast to *Abies*, above).

4 Serbian Spruce

(*Picea omorika*) is a tall slender tree from the Drina Valley, Yugoslavia, but grown more widely for timber and in northern gardens for its spire-like shape.

5 The Cedars

(*Cedrus*) bear needles in whorls rather than singly. Three cedars are commonly grown in the Mediterranean: Atlas Cedar (*Cedrus atlantica*) with branches ascending, from the Atlas Mountains of Algeria and Morocco but much planted elsewhere, especially the blue-leaved form; Cedar of Lebanon (*C. libani*), with fairly level branches, the majestic garden cedar; and Deodar (*C. deodara*), from the Himalayas, with elegant, drooping, branch tips. The wood is pleasantly aromatic but exploitation, especially of *C. libani*, has greatly reduced cedar forests.

5a

5b

5c

5a. A young cedar tree.
5b. The typical barrel-shaped cedar cones; when ripe they fall apart scattering the seeds, but the central axis remains on the tree.'
5c. The foliage of whorled evergreen leaves.

6

7

6 Aleppo Pine

(*Pinus halepensis*) has slender, bright green leaves up to 15 cm long, with grey twigs the first year. It grows in hot, dry, coastal areas.

7 Maritime Pine

(*Pinus pinaster*) has a reddish-brown trunk and dark foliage of rigid, spiny leaves up to 25 cm long. A western Mediterranean tree, it is often planted on dunes to prevent erosion. Resin is extracted from incisions in the bark.

8 Stone or Umbrella Pine

(*Pinus pinea*) has an umbrella-shaped crown and big rounded cones (*c.* 10 × 10 cm) with large edible seeds. It is often planted near the sea. In contrast to other conifers the needles of Pines (*Pinus*) are in clusters of two, three, five or even eight. The species described here have needles in pairs. (See also p. 12.)

8

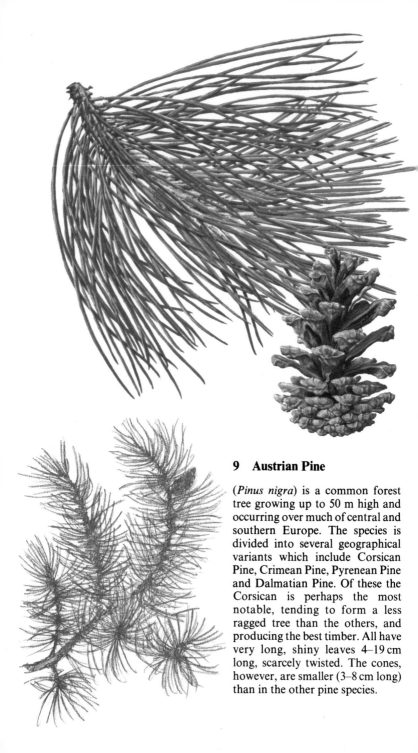

9 Austrian Pine

(*Pinus nigra*) is a common forest tree growing up to 50 m high and occurring over much of central and southern Europe. The species is divided into several geographical variants which include Corsican Pine, Crimean Pine, Pyrenean Pine and Dalmatian Pine. Of these the Corsican is perhaps the most notable, tending to form a less ragged tree than the others, and producing the best timber. All have very long, shiny leaves 4–19 cm long, scarcely twisted. The cones, however, are smaller (3–8 cm long) than in the other pine species.

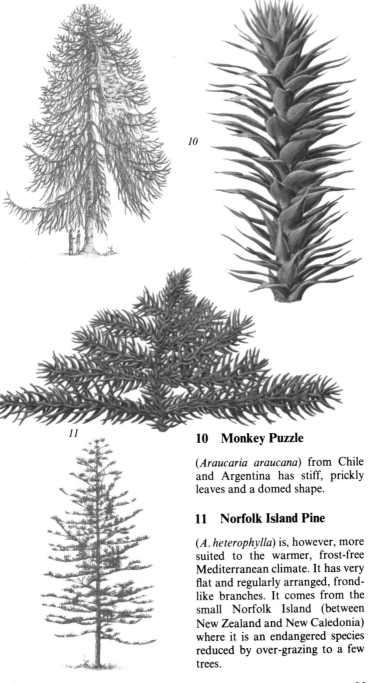

10 Monkey Puzzle

(*Araucaria araucana*) from Chile and Argentina has stiff, prickly leaves and a domed shape.

11 Norfolk Island Pine

(*A. heterophylla*) is, however, more suited to the warmer, frost-free Mediterranean climate. It has very flat and regularly arranged, frond-like branches. It comes from the small Norfolk Island (between New Zealand and New Caledonia) where it is an endangered species reduced by over-grazing to a few trees.

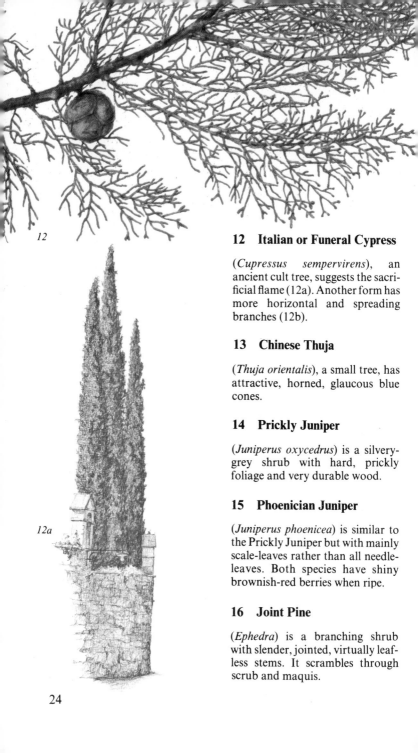

12   Italian or Funeral Cypress

(*Cupressus sempervirens*), an ancient cult tree, suggests the sacrificial flame (12a). Another form has more horizontal and spreading branches (12b).

13   Chinese Thuja

(*Thuja orientalis*), a small tree, has attractive, horned, glaucous blue cones.

14   Prickly Juniper

(*Juniperus oxycedrus*) is a silvery-grey shrub with hard, prickly foliage and very durable wood.

15   Phoenician Juniper

(*Juniperus phoenicea*) is similar to the Prickly Juniper but with mainly scale-leaves rather than all needle-leaves. Both species have shiny brownish-red berries when ripe.

16   Joint Pine

(*Ephedra*) is a branching shrub with slender, jointed, virtually leafless stems. It scrambles through scrub and maquis.

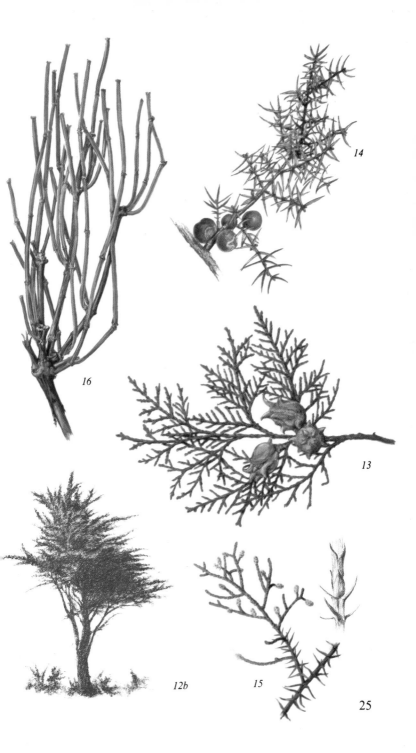

16

14

13

12b

15

25

17   Evergreen Magnolia

(*Magnolia grandiflora*) is a magnificent evergreen tree whose flowers have a deep rich fragrance; in the best forms the leaves are rust-felted underneath, contrasting well with the rich glossy green above. It comes from the warm American states from North Carolina to Texas and Florida, and flourishes in gardens of the less dry parts of the Mediterranean. It is also a good plant for gardens in northern Europe, but in Britain does best planted in a warm corner or against a south wall.

18   Laurel, Sweet Bay

(*Laurus nobilis*) is often seen as a finely clipped, statuesque, small tree with the deep green foliage hiding the black trunk. A native of the Mediterranean, it has been highly esteemed since early times; in ancient Greece it was dedicated

18

to Apollo, and the laurel wreath was presented to victors of athletic contests and to those awarded academic honour. When crushed the leaves exude aromatic oils (try squeezing a leaf) and are an essential herb in good cookery.

19 Virgin's Bower

Two white-flowered species of *Clematis* are common climbing shrubs in the Mediterranean – *C. flammula* with doubly pinnate, often greyish leaves and erect fragrant flowers *c.* 2 cm across, and *C. cirrhosa* (not illustrated) with lobed or toothed leaves and larger nodding flowers 4–7 cm across.

20

21

20 Oriental Plane

(*Platanus orientalis*) is the tree under whose copious shade one sits in the village squares of the Mediterranean countries. These trees were planted, but the species also grows wild in the Balkans. It has been cultivated from the Mediterranean to India since ancient times and was a feature of Persian and Mogul gardens. It is fast growing and can withstand heavy pruning, advantages for a tree planted in towns and villages. The peeling, flaky bark and large, deeply cut leaves are characteristic of the Plane; the flowers are small, in tight heads like tassels on a cord.

21 London Plane

(*P. hybrida*) is thought to be a hybrid between the Oriental Plane and the American Plane (*P. occidentalis*). It is grown as a much-valued street tree in most of Europe, except the drier parts of the Mediterranean, and can be distinguished from the Oriental Plane by its more shallowly cut leaves and flower-heads which usually grow in pairs rather than in groups of three to six.

20

22 Southern Nettle Tree

(*Celtis australis*), native to our area, grows up to 25 m high, with a smooth greyish bark and deciduous leaves like an elm. Its small fleshy berries are sweet and nearly black when ripe.

23 Common or Black Mulberry

(*Morus nigra*) from central Asia is widely cultivated in Mediterranean countries for its edible berries, which are ripe when dark purplish-red. They are very acid until then. It is a deciduous shrub or small rounded tree with leaves varying from ovate to two- or three-lobed.

24 White Mulberry

(*Morus alba*) differs in its hairless, rather than slightly hairy, under-side of the leaf, and longer-stalked berries. A native of China, it is widely naturalized in Mediter-ranean countries and planted both for its berries, which are sweet well before they are ripe, and the leaves upon which silkworms are fed. Silkworms (24a) were cul-tivated in China from antiquity, but how the silk was produced remained a state secret. In *c.* A.D. 550, Emperor Justinian persuaded two Persian monks to smuggle a few silkworms out of China. The silkworm and its fabulous product flourished in medieval Europe.

22

30

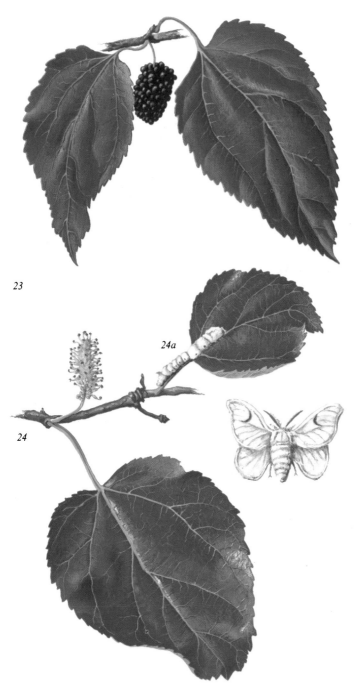

23

24a

24

31

25 Common Fig

(*Ficus carica*) is a small deciduous tree which has been grown in the Holy Land for over 5,000 years and has long been an important element in the diet. Today it is a major crop and one of the principal fruits of the Mediterranean. Since figs remain fresh for a very short time, they are usually stored and exported in the dried state.

26 Indian Rubber Tree

(*Ficus elastica*) is mainly known in Britain as an elegant house-plant. In the Mediterranean it is occasionally seen in town gardens as a large, rounded and very handsome tree.

27 Sycomore Fig

(*Ficus sycomorus*) from East Africa is cultivated in Egypt, but more in the past than today, and is the 'sycomore' of the Bible. The small figs are edible, but insipid compared to those of the Common Fig. The durable wood was used for mummy cases in ancient Egypt (27a).

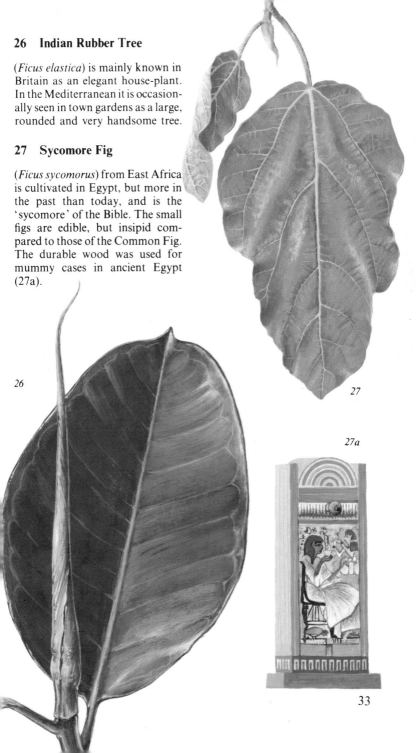

26

27

27a

33

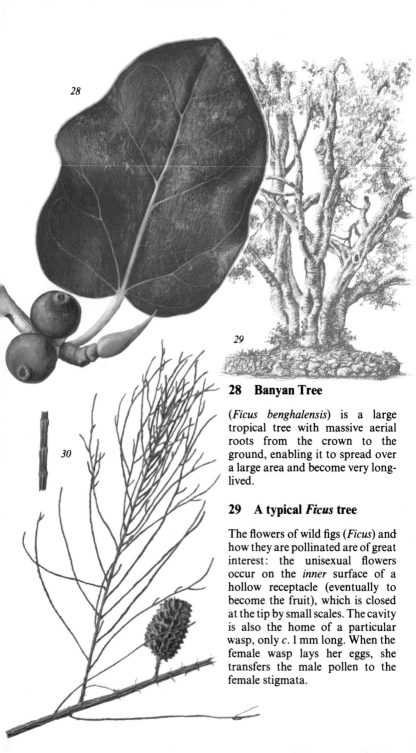

28 Banyan Tree

(*Ficus benghalensis*) is a large tropical tree with massive aerial roots from the crown to the ground, enabling it to spread over a large area and become very long-lived.

29 A typical *Ficus* tree

The flowers of wild figs (*Ficus*) and how they are pollinated are of great interest: the unisexual flowers occur on the *inner* surface of a hollow receptacle (eventually to become the fruit), which is closed at the tip by small scales. The cavity is also the home of a particular wasp, only *c*. 1 mm long. When the female wasp lays her eggs, she transfers the male pollen to the female stigmata.

30 *Casuarina* species are often known as She-oaks or Sheokes and come mostly from Asia and Australia. The virtually leafless, wiry, segmented branches superficially resemble a horse-tail (*Equisetum*) or Joint Pine, but bear minute flowers.

31 Sweet or Spanish Chestnut

(*Castanea sativa*) is a large, deciduous tree which forms woodlands, often in the mountains, from Italy eastwards and north to Hungary. It is much planted and naturalized elsewhere, both as a forest and park tree growing up to 35 m high. The chestnuts themselves are edible and are usually roasted; they are used in particular for flour and in confectionery (*Marrons glacés*).

31

32 Cork Oak

(*Quercus suber*) has very thick, soft, fissured bark which is used to make bottle corks. The first crop of bark is taken when the tree is about 15 years old, but is of little value and is used in tanning and for making cork boards. From then on, the bark can be removed about every 10 or 20 years, and provides cork of good quality. The tree will continue cropping for about 150 years. Cork Oak occurs westwards from Yugoslavia, and is abundant in Spain and Portugal.

33 Holm Oak, Evergreen Oak

(*Quercus ilex*) is a dark, heavy-looking tree found throughout the Mediterranean. It provides good, long-lasting timber.

34 Kermes or Holly Oak

(*Quercus coccifera*) has small spiny leaves resembling those of holly, but acorns typical of an oak. It is often the dominant shrub in the maquis. On it can be found the galls of the scale insect from which ancient peoples made their magnificent scarlet dye.

35 Turkey Oak

(*Quercus cerris*) is deciduous, unlike the other oaks shown here. A large, stately tree, it grows wild from southern and south-central Europe to Turkey.

32

36

38

36 Bougainvillea

(*Bougainvillea*) is one of the most widely planted and conspicuous climbers in subtropical gardens. It covers pergolas and walls with strong red, orange and purple colours. The individual flowers are small and pale yellow, in clusters of threes; it is the brightly coloured bracts surrounding the flowers that provide the colourful effect. Bougainvillea was introduced into Europe from Brazil in 1829.

37 Prickly Pear

(*Opuntia ficus-indica*) is the commonest cactus in the Mediterranean and is widely naturalized; indeed it is a pest in the drier parts of the Canaries and elsewhere, displacing the less vigorous native plants. The true leaves fall very early on, and the swollen, flattened

parts that make up the bulk of the plant are in fact modified stems. It can be propagated by pushing a small piece into the soil which quickly grows into a thick dense hedge. The edible fruits are ovoid and 5–9 cm long; they are sold in markets and are said to be nutritious. The plant was presumably introduced from the Americas, possibly by Columbus.

37

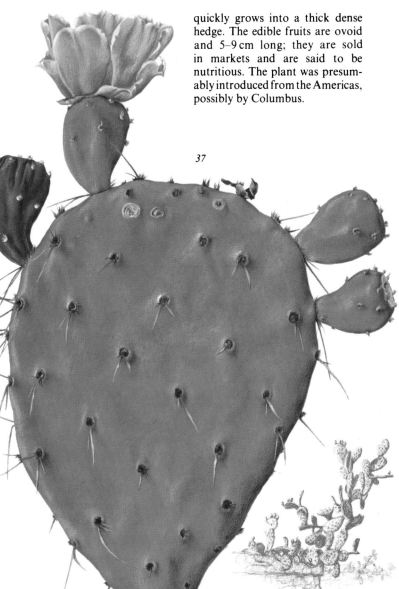

38

38 Plumbago

(*Plumbago capensis*) from the plant-rich Cape of South Africa, is one of the most attractive garden plants of the Mediterranean. Its delicate, powder-blue flowers climb over pergolas, stone walls and archways.

39 Common Camellia

(*Camellia japonica*) is an evergreen shrub or small tree which bears a wealth of blooms up to 10 cm across. Flowers can be single or double, varying from white through pink to pure red. The plant originally grew wild in Japan and Korea and has long been grown as a garden plant in the Orient; it is now widespread in gardens throughout the temperate zone. Most of the forms seen today are the results of many years of breeding and selecting by gardeners for enhanced size, shape and colour of flower.

39

39

39

41

40

40 *Hibiscus rosa-sinensis* is unknown in the wild but is cultivated throughout the tropics and subtropics, making a shrub 1–4 m high. It is also grown as a greenhouse plant in countries such as Britain. There are over 1,000 different cultivated varieties, with single or double flowers, in numerous colours.

41 *Hibiscus syriacus* is also much grown in gardens. The leaves are three-lobed and the flowers white, red, purple or bluish, but usually with a crimson spot at the base.

42　Blue Passion Flower

(*Passiflora caerulea*) is a popular
garden climber, introduced from
Peru and Brazil. The name Passion
Flower was given to it by early
missionaries in South America
who saw a resemblance in the parts
of the flower to the implements of
the Crucifixion; supposedly the
three styles to the nails, the five
stamens to the hammers, and the
fringed corona to the crown of
thorns. Several species of *Passiflora* have edible fruits, such as the
water-lemon and the passionfruit.

43

43

44

44

Sun Roses

(*Cistus*) are a group of about 20 conspicuous and colourful shrubs that grow wild in the maquis, often with Tree Heather, Myrtle and Lentisc, and as the understorey in dry, open pine woods. The leaves are in opposite pairs and are often sticky, in some species exuding a fragrant gum, 'ladanum', which is used in perfumery and medicine. The flowers have five brightly-coloured but short-lived petals and a central boss of numerous yellow or orange stamens. All make good garden plants if given a hot dry position, and will flower abundantly over a long season. Three species are illustrated here:

43 *Cistus monspeliensis* is an aromatic, sticky bush up to 1 m high with white petals and lance-shaped to linear leaves, green above, but densely felted below.

44 *Cistus albidus* has less narrow leaves, densely grey-felted on both sides, and purple flowers. It is barely aromatic.

45 *Cistus ladanifer* is larger than the other members of the genus, reaching $2\frac{1}{2}$ m high. The flowers are solitary rather than clustered, large (up to 10 cm across), either with or without the superb crimson marking at the base of each petal. The branches and lance-shaped leaves are sticky and aromatic.

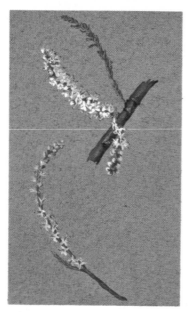

47

46 Tamarisks

(*Tamarix*) are a group of heather-like shrubs or small trees, all very similar to each other and difficult to tell apart. The slender, feathery branchlets bear tiny scale-like leaves, sheathing the stem. Flowers are pink or white, small and delicate but very numerous, in dense spikes that are most attractive in spring. Tamarisks grow all round the Mediterranean Sea, often just at the head of the beach, since they are tolerant of salt water, forming a valuable wind-break.

47 Lombardy Poplar

(*Populus nigra* cv. 'Italica') is a variant of the Black Poplar. It is much used as an ornamental tree for its narrow form and rapid growth. The diamond-shaped leaf is characteristic of all the races of Black Poplar.

47

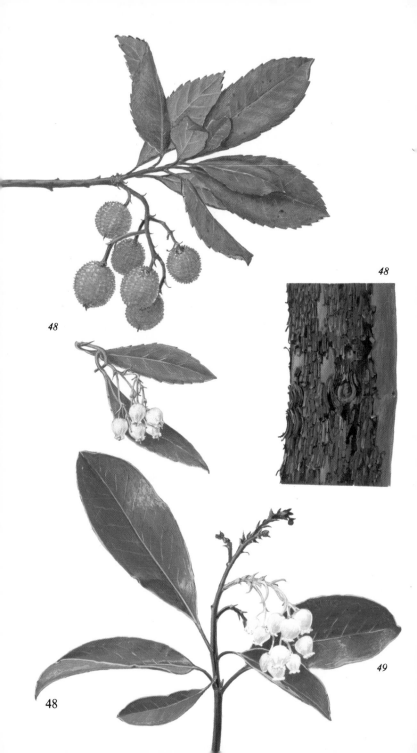

48

48

48

49

48 Strawberry Tree

(*Arbutus unedo*) is an attractive, small tree with dark green, glossy, toothed leaves. It belongs to the heather family, as is shown by the small, urn-shaped flowers. It is a common plant of evergreen scrub and rocky slopes in the less dry areas, occurring in most of the Mediterranean countries and extending locally to sheltered parts of western Ireland. The strawberry-coloured berries are 2 cm across, ripening to deep crimson; they are edible but rather sour, and the name *unedo* implies 'I eat only one'.

49 Eastern Strawberry Tree

(*Arbutus andrachne*), from Albania and Greece to Israel and the Crimea, has a lovely reddish, peeling bark, smooth rather than fissured (as in the Strawberry Tree), and leaves usually untoothed, except where the margin has been eaten by insects.

50 Tree Heather

(*Erica arborea*) grows in similar habitats, forming a thick shrub, but under suitable conditions reaching tree size. The knots formed at the junction between stem and roots are used in briar tobacco pipes because the acid content of the wood makes it almost impossible to burn.

50

49

51

52

50

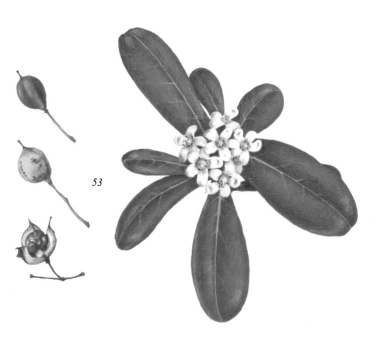

53

51 Kakee or Chinese Persimmon

(*Diospyros kaki*) is a small deciduous tree with a rounded crown, introduced from China and Japan where it is much grown for its fruits. The flowers are yellowish-white and *c.* 4 cm across; the fruits are large, reaching the size of a peach, but are shaped more like a tomato. Since they are barely edible until they are over-ripe, they are harvested late in the year, after the leaves have fallen.

52 Date Plum

(*Diospyros lotus*) has fruits that are yellowish or purplish, with a dew-like bloom, and are no larger than a cherry; they taste sweet but insipid.

53 *Pittosporum tobira* is easy to recognize with its shining, dark green leaves and strongly scented flowers, first white then yellow. It is used extensively in parks and gardens, and can withstand frequent cutting so that although it can reach tree size, it is more usually seen as a low, clipped shrub, often in a hedge.

53

54

55

56

54 Almond

(*Prunus dulcis*) probably came from the Near East, but is now so widely cultivated it is difficult to trace. It is a small tree, often tending to become spiny and intricately branched if not pruned. It is grown in the Mediterranean for the nuts, and often becomes naturalized. It is also grown in gardens further north for its early flowers. The petals open as bright pink but fade to white. The fruit is green, dry and leathery; inside the outer layer the stone is deeply wrinkled and when cracked reveals the oil-rich seed – the almond.

55 Peach

(*Prunus persica*) came from China but is now widely cultivated in the subtropics and where there are warm summers. As with all plants of ancient culture, there are innumerable variants, in colour, shape and flavour. The smaller but more aromatic form with completely smooth rather than velvet-haired skin is the nectarine.

56 Apricot

(*Prunus armeniaca*) forms a taller tree than either peach or almond, and its leaves are broad, almost heart-shaped, on long stalks. The flowers are pure white or of the palest pink; the stone of the fruit is smooth rather than fissured.

All three plants flower early in the Mediterranean; the almond comes first, before the leaves unfold, but the others quickly follow. At blossom time the orchards are a delight to the senses.

57

58

57 Portugal Laurel

(*Prunus lusitanica*) is an evergreen that grows into a hedge plant or small tree. In the wild it occurs from Portugal to France, but is often planted in other parts of western Europe.

58 Cherry Laurel

(*Prunus laurocerasus*) is similar but has young twigs and leaf stalks which are green rather than bright red. It grows wild in the Balkan Peninsula, but again is widely planted in parks and gardens of

southern and western Europe. Both laurels are hardy, but it is the Cherry Laurel that is common in English gardens and has become naturalized in parts of the country.

59 Loquat

(*Eriobotrya japonica*) is an evergreen shrub or small tree with large coarse leaves, the undersides of which are covered with brown felt-like hairs, as are the branches. The fruit, which is about the size of a small plum, is very juicy and sweet, containing a few large seeds. It has

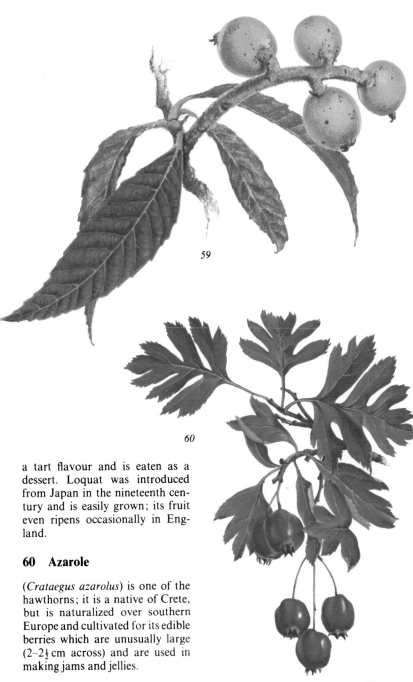

59

60

a tart flavour and is eaten as a dessert. Loquat was introduced from Japan in the nineteenth century and is easily grown; its fruit even ripens occasionally in England.

60 Azarole

(*Crataegus azarolus*) is one of the hawthorns; it is a native of Crete, but is naturalized over southern Europe and cultivated for its edible berries which are unusually large (2–2½ cm across) and are used in making jams and jellies.

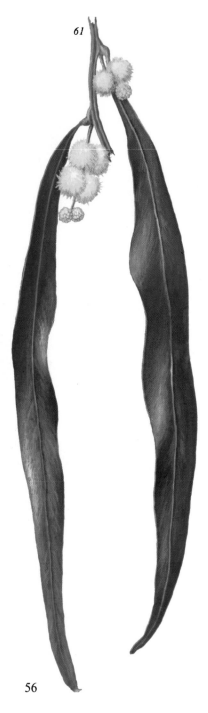

61

Acacia, Wattle

(*Acacia*). The acacias are one of the largest groups in the plant kingdom with around 1,000 species, mostly from tropical Africa and Australia. The African species tend to be the low, flat-topped trees of the grassland savannahs. Acacias are easily recognized by their attractive and very striking, bright yellow flower-heads which are produced in abundance. The individual flowers are small and insignificant; the bright yellow effect comes from the stamens which protrude from the numerous flowers making up each small head. Some acacias (e.g. 62) have doubly pinnate leaves; in others (e.g. 61) the pinnate leaf is replaced by a flattened and widened leaf stalk that serves as a simple leaf. Some have tough thorns. Ants bore through the thorns of some species to make their homes. Many acacias are useful to man: *A. senegal* is the source of the best gum arabic, exuded from the branches in dry desert winds. Some have beautiful timbers and the two species included here are important ornamentals.

61   Blue-leaved Wattle

(*Acacia cyanophylla*) forms a small, slender tree, with pendulous twigs bearing heads of strongly scented flowers.

62   Silver Wattle, Mimosa

(*Acacia dealbata*) forms a silvery-grey, evergreen tree with long racemes of deliciously scented flower-heads. It is very common in Mediterranean gardens, valued for its early flowering, and is the 'mimosa' of florists.

56

63 Persian Acacia, Pink Siris

(*Albizia julibrissin*) is similar to an acacia, but in this species the stamens are very long and reddish, like silken threads.

64 Barbados Pride

(*Caesalpinia pulcherrima*) is a prickly shrub about 3–4 m high, a feature in many tropical and subtropical parks and gardens.

65 Flamboyant

(*Delonix regia*) is a magnificent ornamental tree up to 15 m high, with a spreading crown

66

and abundant flowers on bare branches. Although widely grown in tropical and subtropical gardens, it grows in the wild only in one small forest in Madagascar, where it is threatened by clearing and burning.

66 Judas Tree

(*Cercis siliquastrum*) is a small, rounded tree often cultivated but also growing wild in the Mediterranean. It has abundant pink to purple flowers, usually on the bare twigs, and is the tree on which Judas is traditionally believed to have hanged himself; after this the flowers, then reputedly white, blushed forever in shame.

67

68

60

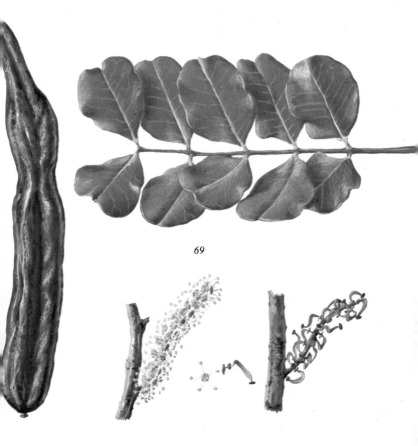

69

67–68   Cassia

(*Cassia*). There are about 500–600 species of *Cassia* from the tropics and warm temperate areas (excluding Europe). They are the source of the drug senna which is obtained from drying the leaves or the pulp of the pods. Others are good ornamentals: *Cassia didymobotrya* (67) from tropical Africa, and *Cassia corymbosa* (68) from the New World.

69   Carob, Locust Tree

(*Ceratonia siliqua*) is both native and widely cultivated in the Medi-

terranean. It is a small, rounded tree with coarse foliage and small, inconspicuous flowers which have an unpleasant smell, particularly at night. When ripe the pods contain a sweet pulp of sugar and gum, with several large seeds; they are used as cattle fodder, but were formerly human food, and it has been suggested John the Baptist lived on them in the desert. The seeds are relatively uniform in size and were the original 'carat' of jewellers; since, however, the original carat did vary from one place to another, the carat weight has now been standardized at 0.200 gm.

70

70   Honey Locust

(*Gleditsia triacanthos*) is a tall tree
introduced from North America;
it is easily recognized by the
formidable clusters of branched
spines on the trunk. It is grown
both for its foliage, which turns
a fine yellow in the autumn, and
for its numerous, twisted pods
which remain on the tree in winter,
hanging down and rustling in the
wind.

71   Pagoda or Scholar's Tree

(*Sophora japonica*) from China has
creamy white, pea-shaped flowers,
which appear only on old trees and
then late in the year. It is famous
for its wrinkled and contorted
trunk and branches. The Locust
Tree or False Acacia (*Robinia
pseudacacia*) is similar but has a
characteristic pair of spines at the
base of each leaf.

All the species on this page are
vigorous and hardy and are grown
in English gardens as well as in the
Mediterranean. They make handsome trees up to 30 m or more high.

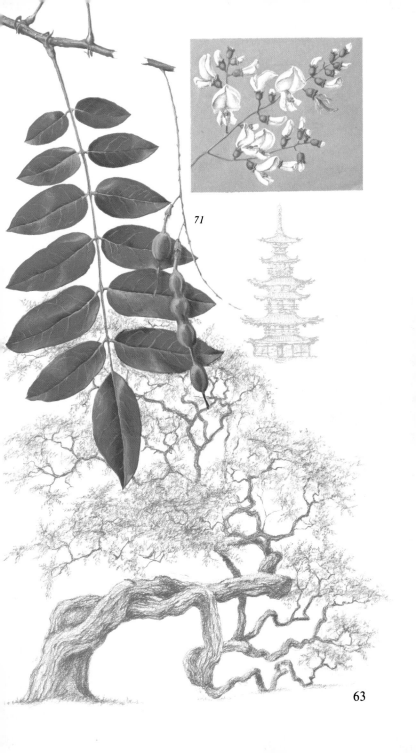

71

72

73

74

64

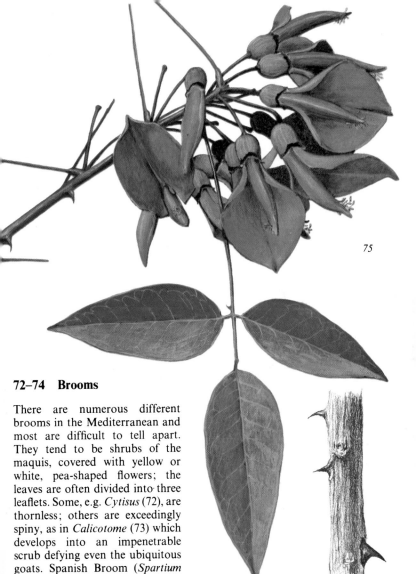

72–74   Brooms

There are numerous different brooms in the Mediterranean and most are difficult to tell apart. They tend to be shrubs of the maquis, covered with yellow or white, pea-shaped flowers; the leaves are often divided into three leaflets. Some, e.g. *Cytisus* (72), are thornless; others are exceedingly spiny, as in *Calicotome* (73) which develops into an impenetrable scrub defying even the ubiquitous goats. Spanish Broom (*Spartium junceum*) (74) is, however, easily recognized by its long, smooth, distinctly rush-like branches that are used for basket-making. It is perhaps the most colourful of the brooms – a bush of blazing yellow by the roadside where it is so often planted.

75   Coral Tree

(*Erythrina crista-galli*) from Brazil is widely grown in gardens for its magnificent, bright scarlet heads of flowers.

76

76 Crape Myrtle

(*Lagerstroemia indica*) is a colour-ful shrub or small tree, grown for the mottled silky sheen of the bark and the attractive flowers whose petals are distinctively crinkled.

77–78 Eucalypts, Gums

(*Eucalyptus*) is a large and easily recognizable group of about 500 species belonging to the myrtle family. The petals are joined to-gether to form a cap over the bud, which when the flower opens is pushed off by the numerous stamens as they unfold. Many species have two types of leaves, those on the young saplings tend-ing to be broader, glaucous and horizontal, whereas those on mature trees are narrower, greener and pendulous, as shown opposite (77) for Tasmanian Blue Gum. Eucalypts come from Australia where most are tall trees of open savannah or woodland. About fif-teen species have been introduced to the Mediterranean and are now common and conspicuous trees in the landscape, planted both for forestry and along avenues. They grow extremely fast and flourish on the less good soils. Because of their great intake of water, they have been planted in marshes, breeding areas of the malarial mosquito, in part to dry out the habitat and so help to eradicate the disease. Shown here are Tas-manian Blue Gum (*Eucalyptus globulus*) (77), a tall, most attrac-tive tree up to 40 m high, widely planted in parks and along avenues; and Swamp Mahogany (*Eucalyptus robustus*) (78) which grows in damp areas, tolerating slightly saline soils.

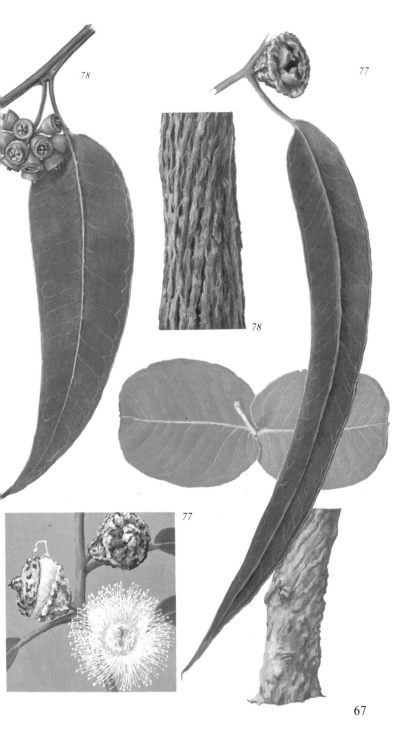

78

77

78

77

67

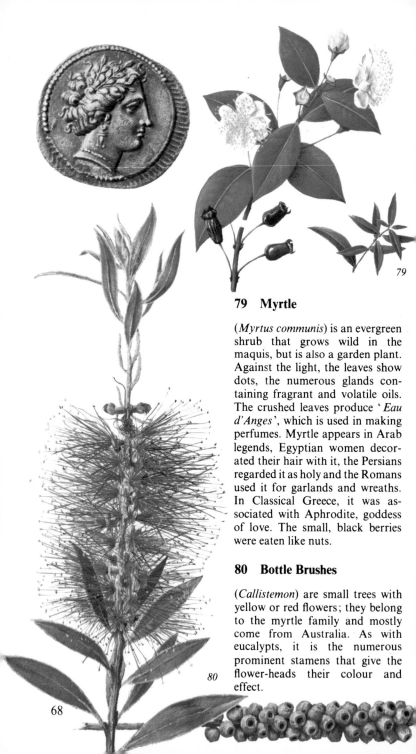

79 Myrtle

(*Myrtus communis*) is an evergreen shrub that grows wild in the maquis, but is also a garden plant. Against the light, the leaves show dots, the numerous glands containing fragrant and volatile oils. The crushed leaves produce '*Eau d'Anges*', which is used in making perfumes. Myrtle appears in Arab legends, Egyptian women decorated their hair with it, the Persians regarded it as holy and the Romans used it for garlands and wreaths. In Classical Greece, it was associated with Aphrodite, goddess of love. The small, black berries were eaten like nuts.

80 Bottle Brushes

(*Callistemon*) are small trees with yellow or red flowers; they belong to the myrtle family and mostly come from Australia. As with eucalypts, it is the numerous prominent stamens that give the flower-heads their colour and effect.

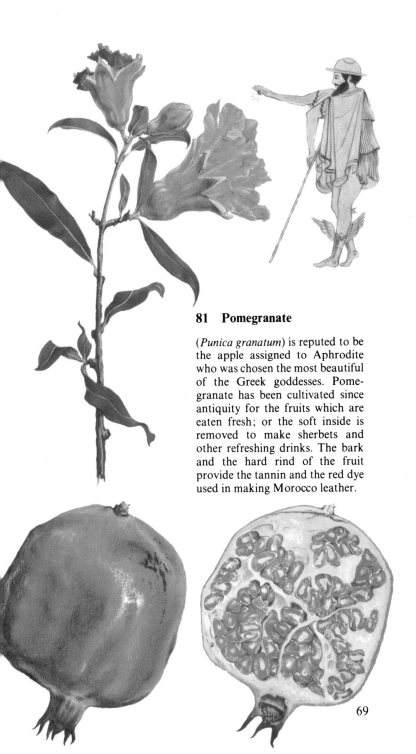

81 Pomegranate

(*Punica granatum*) is reputed to be the apple assigned to Aphrodite who was chosen the most beautiful of the Greek goddesses. Pomegranate has been cultivated since antiquity for the fruits which are eaten fresh; or the soft inside is removed to make sherbets and other refreshing drinks. The bark and the hard rind of the fruit provide the tannin and the red dye used in making Morocco leather.

69

82 Oleaster

(*Eleagnus angustifolia*) is an attractive but rather spiny shrub or small tree, whose grey, willow-like leaves are covered on the underside by minute silvery scales. The small yellow berries are succulent and sweet to the taste. Oleaster, from temperate parts of Asia, is planted as an ornament and for its berries; it has also become widely naturalized.

83 Aucuba, Spotted Laurel

(*Aucuba japonica*) is an evergreen bush from Japan. In some forms the coarse foliage is spotted or variegated. Its drab appearance is enlivened only by the clusters of small, scarlet berries. The flowers are unisexual, with males and females on separate plants.

84 Castor Oil Plant

(*Ricinus communis*). This is a tropical plant which has been cultivated in Egypt since about 4000 B.C. In frost-free areas it makes a small tree, but otherwise can be grown as a summer annual. The large, marbled seeds are remarkably rich in oil, which is removed by pressing and is used in medicine, in oil lamps and as a lubricant. The seeds contain ricin which is highly poisonous, but the poison remains in the pressed pulp.

71

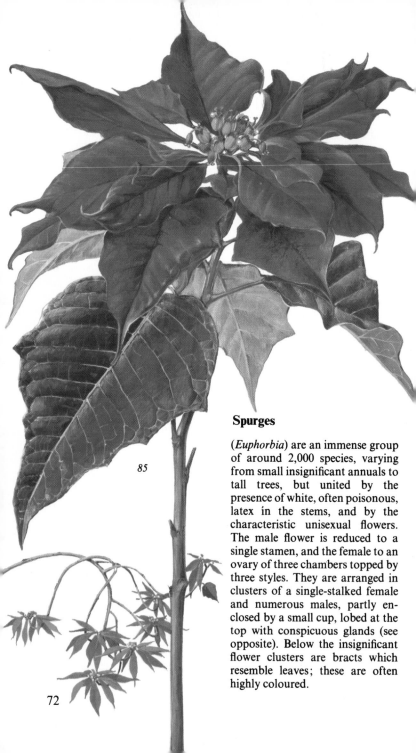

85

Spurges

(*Euphorbia*) are an immense group of around 2,000 species, varying from small insignificant annuals to tall trees, but united by the presence of white, often poisonous, latex in the stems, and by the characteristic unisexual flowers. The male flower is reduced to a single stamen, and the female to an ovary of three chambers topped by three styles. They are arranged in clusters of a single-stalked female and numerous males, partly enclosed by a small cup, lobed at the top with conspicuous glands (see opposite). Below the insignificant flower clusters are bracts which resemble leaves; these are often highly coloured.

72

86

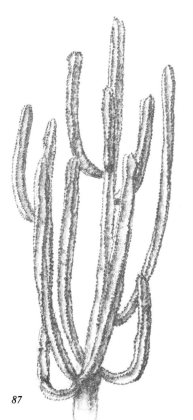

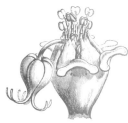

85 Poinsettia

(*Euphorbia pulcherrima*) is a familiar pot-plant which in the subtropics grows into a small tree. Both the young leaves and the bracts below the flowers are highly coloured.

86 Tree Spurge

(*Euphorbia dendroides*) is a woody shrub up to 2 m high, common on rocky places overlooking the sea.

87 *Euphorbia canariensis* has leafless, succulent, spiny stems resembling those of a cactus. It grows wild in the dry coastal region of the Canaries, but is also grown elsewhere as an ornamental.

87

88 Mediterranean Buckthorn

(*Rhamnus alaternus*) is an evergreen bush with a wood that smells coarse and unpleasant. The small yellowish-green flowers are either male or female, but both usually occur on the same plant; from the females develop the berries which ripen through red to black.

88

89 Christ's Thorn

(*Paliurus spina-christi*) is a very strongly thorned shrub which grows into dense tangles, especially in the drier parts of the Mediterranean. It is often used as a hedge plant. There are two characteristic thorns at the base of each pair of leaves on the zig-zagging stems; one is straight, the other hooked and curving downwards. Like so many plants of the eastern Mediterranean which have thorny twigs, it has been suggested that it is the plant used to make the Crown of Thorns.

89

90   Common Jujube

(*Ziziphus jujuba*) resembles *Paliurus* but the ripe fruits are soft, sweet and tasty rather than membranous and dry; they become dark red to almost black when ripe, and about the size of olives. From ancient times they have been used for coughs. The plant is widely cultivated and naturalized, coming originally from Asia. Another species, *Ziziphus lotus*, has grey rather than green twigs and a deep yellow, spherical berry; it is a native of the southern Mediterranean and in the past was a source of human food; it was reputed to be the lotus fruit of antiquity.

91 Vines, Grapes and Wine

The vine (*Vitis vinifera*) is a climber with a woody stem and shoots that climb and twine with the help of their tendrils. The fruit is a soft berry, green, yellow, red or black, with a particularly high sugar content. The original homeland of the vine seems to have been Asia, but the plant has been in cultivation from a very early date – certainly from 4000 B.C., and probably earlier, so wine and hence alcohol have been familiar to man since the earliest civilizations. Over the centuries the centre of vine culture has moved steadily westwards, from Mesopotamia to ancient Egypt and to Greece, and in Roman times to Germany, Italy and France. Today the vine is grown in most warm-temperate and subtropical regions, especially in Australia, California and South Africa as well as the Mediterranean basin. One of the secrets of the grape is its sweetness; without adding any sugar, the juice will ferment into a liquid with ten to fifteen per cent alcohol. In general the hotter the summer, the sweeter the grape and so the higher the alcohol content. Whereas white wine is made from the juice obtained from pressing the grapes, red wine is made from the whole grape, which is usually crushed or squeezed to help the fermentation along. The tannin in the skins plays a vital role in giving the various red wines their unique tastes and aromas.

Grapes are also dried in the sun to produce raisins, in Greece and in Spain particularly, and from the seedless varieties, currants, notably in southern Italy.

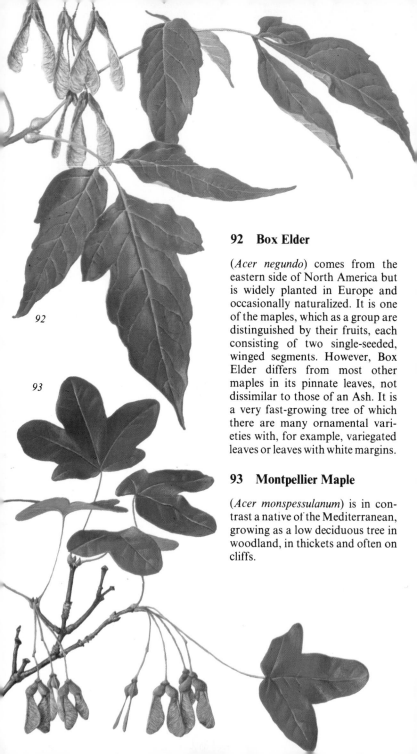

92　Box Elder

(*Acer negundo*) comes from the eastern side of North America but is widely planted in Europe and occasionally naturalized. It is one of the maples, which as a group are distinguished by their fruits, each consisting of two single-seeded, winged segments. However, Box Elder differs from most other maples in its pinnate leaves, not dissimilar to those of an Ash. It is a very fast-growing tree of which there are many ornamental varieties with, for example, variegated leaves or leaves with white margins.

93　Montpellier Maple

(*Acer monspessulanum*) is in contrast a native of the Mediterranean, growing as a low deciduous tree in woodland, in thickets and often on cliffs.

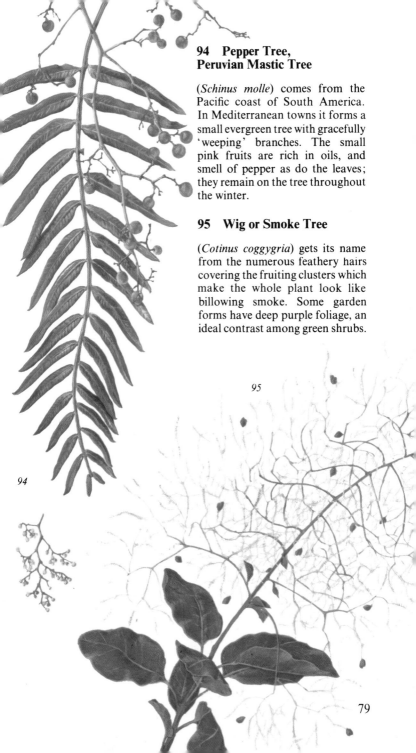

94 Pepper Tree, Peruvian Mastic Tree

(*Schinus molle*) comes from the Pacific coast of South America. In Mediterranean towns it forms a small evergreen tree with gracefully 'weeping' branches. The small pink fruits are rich in oils, and smell of pepper as do the leaves; they remain on the tree throughout the winter.

95 Wig or Smoke Tree

(*Cotinus coggygria*) gets its name from the numerous feathery hairs covering the fruiting clusters which make the whole plant look like billowing smoke. Some garden forms have deep purple foliage, an ideal contrast among green shrubs.

95

94

96

96 Mastic Tree, Lentisc

(*Pistacia lentiscus*) is usually seen as a low evergreen bush, particularly common in maquis, but if left uncut and ungrazed it can reach tree size. The whole plant smells acrid and resinous. An incision on the bark brings a flow of the mastic resin, which is collected when it has dried. The resin has been used since ancient times for medicinal purposes, as a varnish, and as chewing gum to clean the teeth and freshen the breath. The Aegean island of Chios has long been famed as a producer of mastic resin.

98

97

99

97 Terebinth, Turpentine Tree

(*Pistacia terebinthus*) is similar to the Lentisc, but is deciduous and the pinnate leaves include a terminal leaflet. It is, however, just as strong-smelling and produces a gum resin. Both species are found throughout the Mediterranean region.

98 Pistachio

(*Pistacia vera*) differs from the Lentisc and Terebinth by the leaf having only one to three leaflets. The large fruit, *c.* 2½ cm long, has a hard outer shell enclosing edible seeds, the Pistachio nuts. The plant was introduced to the Mediterranean from Asia.

99 Tree of Heaven

(*Ailanthus altissima*) is a tall tree which, although attractive, does not quite live up to its splendid name. It is a native of China, but is widely planted in towns and parks. The bark is smooth and the pinnate leaves are long and striking; they have an unpleasant smell.

100–101 Oranges and Lemons

Citrus fruits originally came from China and southeast Asia. Over the years they have been continually bred for characteristics of value to man, and so they are not only very diverse from each other, but also greatly differ from the original wild forms. All, however, are evergreen shrubs or small trees whose leathery leaves are perforated with oil glands. The white flowers appearing in the leaf axils have a delicious fragrance that fills the air at blossom time. The fruits consist of a thick leathery skin and a succulent interior containing

the juice in numerous swollen hairs.

Many citrus fruits are grown in the Mediterranean, not only for their fruits but also for their flowers which are used in the perfume industry. Citrus orchards are usually planted in the lowlands, often near the sea. Orange (100) and Lemon (101) are the most common, but others are also grown: Seville Orange, similar to Orange, but with a sour, bitter pulp used for making marmalade; Tangerine, with smaller, sweet, orange fruits and thin rind that peels easily; Citron, with very thick, rough and warty yellow rind (the only one with a leaf stalk that is not winged); Shaddock or Pomelo, with giant yellow fruit, up to 25 cm across and the well-known Grapefruit.

102 Indian Bead Tree, Persian Lilac

(*Melia azederach*) is a small tree often planted in towns and along roads. The seeds of the small yellow fruits are sometimes used as the beads in rosaries, hence its English name. Although the flowers are lilac-coloured and fragrant, they have little else in common with Lilac, nor does the plant come from Persia, so the other part of the name is a misnomer.

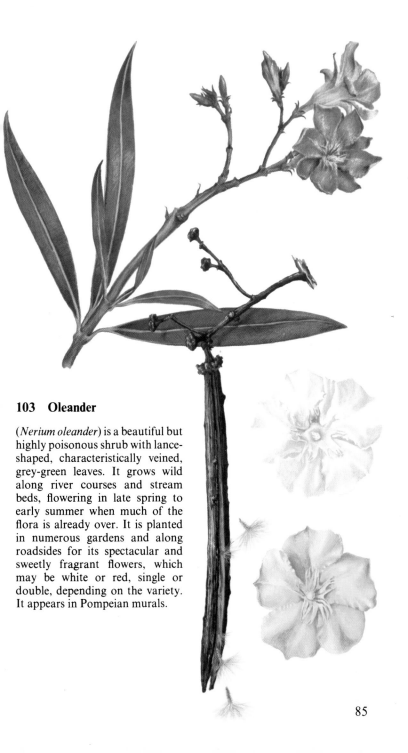

103 Oleander

(*Nerium oleander*) is a beautiful but highly poisonous shrub with lance-shaped, characteristically veined, grey-green leaves. It grows wild along river courses and stream beds, flowering in late spring to early summer when much of the flora is already over. It is planted in numerous gardens and along roadsides for its spectacular and sweetly fragrant flowers, which may be white or red, single or double, depending on the variety. It appears in Pompeian murals.

85

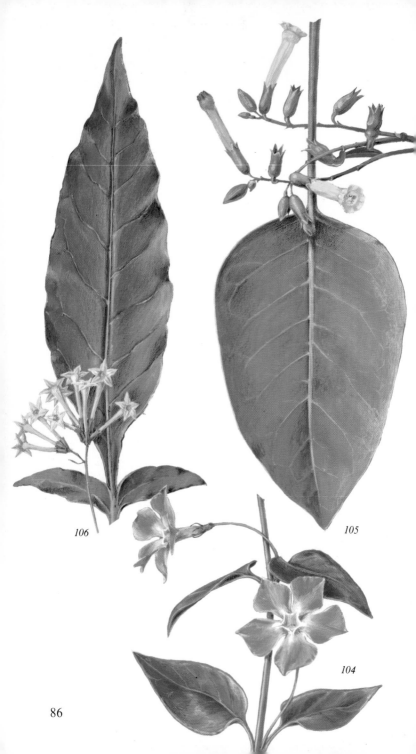

106

105

104

104 Greater Periwinkle

(*Vinca major*) is a low creeper with broad evergreen leaves in pairs and large blue flowers. It grows wild in shaded places, often in woodland, and so makes a good ground-cover under garden shrubs.

105 *Nicotiana glauca* from South America is a slender shrub with yellow tubular flowers and is related to the plant from which tobacco is produced. It is commonly planted and naturalized by roadsides, flowering throughout the year.

106 *Cestrum parqui* from Chile is a deciduous shrub whose yellow flowers are scented at night.

107 Thorn Apples

(*Datura*) contain highly poisonous alkaloids, despite their most spectacular hanging flowers and overpowering scent. Most come from South America.

108

109

108 Chaste Tree

(*Vitex agnus-castus*) is one of many plants that have played a large part in myth and folklore. The small seed has a sharp taste and is sometimes called 'Monk's Pepper', since it is reputed to repress the sexual instinct. Thus the plant was associated with chastity. It has delicate foliage and lilac-blue flowers in long spikes.

109 *Lantana camara* is a colourful and gaudy shrub from the Caribbean. It is often planted for its unusual flowers which, as they age, change from yellow (or pink) to orange or red, so that both colours appear in the same head of flowers.

110 Rosemary

(*Rosmarinus officinalis*) is a common wild plant in dry scrub around the Mediterranean. Its aromatic leaves are a main herb in cookery and an ingredient in cosmetics and perfumes. It was an ancient symbol of fidelity.

110

89

111

112

111–114   Jasmine

Perfume-making is an important industry in some areas of the Mediterranean. Among the many plants producing the basic ingredients are the jasmines, a group well known for the evocative scent of their flowers. In France they are grown in large plantations. The species cultivated include the scrambling *Jasminum officinale* (111) from Asia, and the larger-flowered but similar *J. grandiflorum* (112). Of these, *J. officinale* is the commoner in gardens and is often naturalized. With alternate, rather than opposite, leaves, are the yellow-flowered and very fragrant *J. odoratissimum* (113) from Madeira and the Canary Islands (but cultivated elsewhere), and *J. fruticans* (114), which grows wild around the Mediterranean. These two are shrubs rather than climbers.

113

114

115 The Olive

(*Olea europaea*). 'There are two liquids which are pleasing to the human body, inwardly wine and outwardly oil, and they both come from trees.' So wrote Pliny the Elder about wine and olive oil some two thousand years ago. It is believed that the Olive Tree has been grown ever since the earliest civilizations in the Mediterranean basin. Olive oil is used for cooking and lighting, and has a host of other applications; the olives them-selves are an important element in the diet, often eaten with bread or salad. At a distance an olive tree looks like an ancient willow tree, often gnarled with deeply furrowed bark. The flowers are small, whitish and four-lobed, in dense clusters at the nodes. The fruit is first green, later becoming blackish, and contains a hard stone. Light pressing, when only the fleshy fruit is pressed, produces the finest edible oil, which is strained to remove the bitter solid remains of the fruit.

116

116 Glossy or Chinese Privet

(*Ligustrum lucidum*) is a large ever-green shrub or tree up to 10 m high, from China and Japan. It is often planted as an avenue tree and when in flower it is covered with large and fragrant sprays.

117 Manna or Flowering Ash

(*Fraxinus ornus*) is 8–20 m high, with a rounded crown. It grows wild in central and southern Europe, and occurs mainly in woodland and thickets. It is easily recognized by its conspicuous conical heads of white flowers (117a), appearing when the leaves are still small. A sweet gum called 'Manna' is exuded from the branches and is used in folk medicine.

117

117a

118 *Phillyrea angustifolia* grades closely into the wider-leaved *P. latifolia*; both are common and in places dominant. They are small trees in evergreen scrub and woodland, especially near the coast.

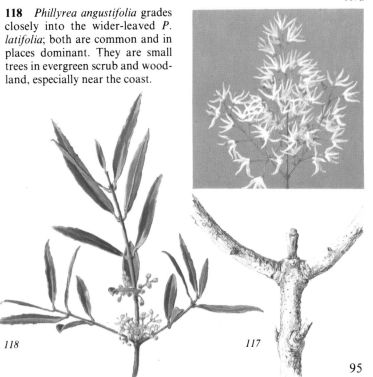

118

117

119

119 Paulownia or Foxglove Tree

(*Paulownia tomentosa*) is a rounded tree with woolly twigs, large foliage and magnificent violet flowers in large conical sprays. It comes from China and is widely planted in Mediterranean parks and gardens, as well as in those of northern Europe.

120 Indian Bean Tree

(*Catalpa bignonioides*) has a similar shape and foliage but the flowers are whitish, speckled yellow and purple, and are in numerous clusters of threes. The fruits are long and slender, hanging down in great quantities (the beans). This tree does not come from India but from the southern part of the United States of America.

120

121–124 Trumpet Creepers

The strongly coloured trumpet climbers of the *Bignonia* family grow on walls in frost-free areas. Cape Honeysuckle (*Tecomaria capensis*) is from South Africa (121), *Pyrostegia venusta* from Brazil and Argentina (122), *Phaedranthus buccinatorius* from Mexico (123) and *Campsis radicans* from Pennsylvania to Texas (124).

121

122

123

124

125

125 Jacaranda

(*Jacaranda mimosifolia*) is another member of the *Bignonia* family. It is widely grown in the tropics and subtropics for its delicate foliage and magnificent pale violet flowers.

126 Gardenia

(*Gardenia jasminoides*) from China and Japan is a shrub with flowers that in the cultivated forms are double and of the purest white. They have an exceptionally fine scent and are most famous as cut flowers.

126

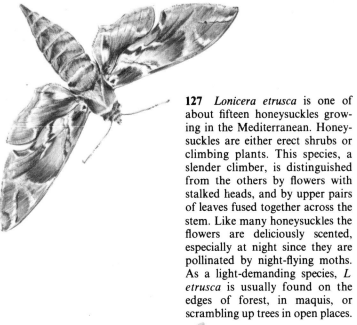

127 *Lonicera etrusca* is one of about fifteen honeysuckles growing in the Mediterranean. Honeysuckles are either erect shrubs or climbing plants. This species, a slender climber, is distinguished from the others by flowers with stalked heads, and by upper pairs of leaves fused together across the stem. Like many honeysuckles the flowers are deliciously scented, especially at night since they are pollinated by night-flying moths. As a light-demanding species, *L etrusca* is usually found on the edges of forest, in maquis, or scrambling up trees in open places.

128 Laurestinus

(*Viburnum tinus*) is a large, rounded
evergreen shrub or small tree, with
dark green, smooth, glossy leaves
somewhat similar to those of laurel,
which accounts for the English
name. Laurestinus grows wild
in southern Europe and North
Africa, and is usually found in
woodlands or in more open sites.
It is much grown further north
in gardens, especially in England,
where it flowers between December
and April, when few other plants
are at their best. The blue fruits
become black as they mature, and
glisten with a metallic sheen.

Palms

One of the most distinctive and remarkable groups in the plant kingdom, palms are very different from the other trees in this book. Their form is amazingly simple: a crown of leaves at the top of a straight, unbranched stem. The leaves are either pinnate (as with the date palms) where the leaf axis elongates as the leaf develops, or fan-shaped (as in *Chamaerops*, opposite). There is only one apex on the whole tree, in the middle of the crown; if it is broken or removed, the stem dies. When a palm is only about one metre high, its trunk will have usually reached the thickness it will retain to the end of its life. Thus, unlike oaks or pines, there are no annual growth rings. As for the flowers, they are small and massed together in very large, much-branched panicles, often hidden before bud-burst in a sheath. The palm family includes 2,500–3,000 species, but most are plants of the wet tropics. Many have a host of economic uses: trunks for timber, sap for food and leaves and fruits for numerous other uses. In some societies a single species of palm can supply all the necessities of life.

129 Dwarf Fan Palm

(*Chamaerops humilis*) is one of the two palms growing wild in southern Europe. It occurs in North Africa and from Portugal to Italy, but is widely grown elsewhere in gardens where it will reach 4 m high; in the wild it often cannot develop a trunk because of continual grazing by sheep and goats, and it becomes low and densely clustered. The rough and coarse leaf fibre is made into stuffing for furniture and mattresses, a 'vegetable horse hair'.

130 Date Palm

(*Phoenix dactylifera*) is one of the oldest of cultivated plants; it was widely grown in the Egypt of the Pharaohs, when every part was utilized. Dates ripen only in a very hot and dry climate, and thus Date Palms flourish in the Middle East where they are essential to life – and also provide a good export product. Dates and milk are reputed to be an ideal diet in the desert and the lower grades of dates are a good cattle fodder. But dates are not the only product of this remarkable plant: the prickly leaf stalks can be used for fences, the young leaf fronds for roofing, the leaflets for making mats and baskets, the fibre of old leaves for twisting into rope, the sap for a palm wine and the wood for planks. Palm leaves have long been associated with religious festivals, in particular Palm Sunday.

131 Canary Date Palm

(*Phoenix canariensis*) grows wild in the Canaries, but is the most widely-used ornamental palm in the northern Mediterranean, especially as a street tree near the coast, since it can withstand cool temperatures and sea winds. It differs from the Date Palm in its stouter, shorter trunk, dry, almost inedible fruits and larger, leafier crown.

132 *Washingtonia filifera* is a fan-leaved palm from California, growing up to 15 m in cultivation. Its specific name (*filifera*) refers to the long threads which hang between the leaflets. Another characteristic feature is that the dead leaves tend not to fall, but hang like a skirt, hiding part or all of the trunk.

133 Doum Palm

(*Hyphaene thebaica*) is one of the few exceptions to the rule that palms do not have branches. It is easily recognizable in pictures from ancient Egypt where it has been cultivated since time immemorial. The fruits are sold both as a food and as a drug; they have a sweet, spicy taste. The stone-hard kernels in the fruit are sometimes used as 'vegetable ivory', to make buttons and small wood carvings. The Doum Palm provides a good timber and the young leaves can be eaten by camels.

134 *Livistona* is a group of some of the best ornamental palms, with tall trunks 15–25 m high and shapely crowns of graceful fan leaves. The leaf bases are very tough and fibrous. The illustration is of *Livistona australis* from eastern Australia, reputed to be the hardiest member of the group.

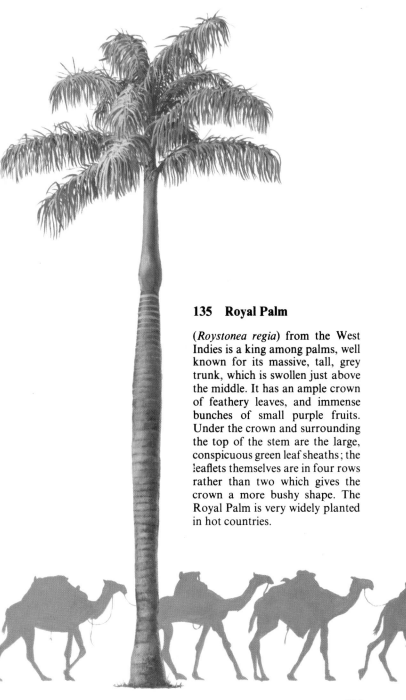

135 Royal Palm

(*Roystonea regia*) from the West Indies is a king among palms, well known for its massive, tall, grey trunk, which is swollen just above the middle. It has an ample crown of feathery leaves, and immense bunches of small purple fruits. Under the crown and surrounding the top of the stem are the large, conspicuous green leaf sheaths; the leaflets themselves are in four rows rather than two which gives the crown a more bushy shape. The Royal Palm is very widely planted in hot countries.

111

136 Canarian Dragon Tree

(*Dracaena draco*) from the Canaries and Madeira is commonly grown in the Mediterranean because of its distinctive and most peculiar appearance. The multi-branched stems end in a tight cluster of sword-shaped leaves up to one metre long. From the trunk exudes a red gum, called 'Dragon's Blood', which was in great demand in the past because of its reputed medicinal and magical properties. The plant is said to reach great age, but few old trees in the wild have survived felling. The largest and most famous, which was said to be 78 feet in diameter and over 6,000 years old, died in 1867.

Trees in History: in Religion, Mythology and Design

When man first colonized the Mediterranean basin the forest cover was extensive and trees played an important part in sustaining human life. From the forest the essentials of life were obtained – the meat of wild animals and birds, berries, nuts and edible roots. He gathered materials to build shelter and make implements and tools. His economy was based on the direct use of the available natural resources. The forest was hazardous, however, with predatory animals and poisonous plants. The life of the hunter and food-gatherer was physically and mentally dangerous, and it is not surprising that individual trees of peculiar shapes, mysterious phenomena like the fall of the leaves, the flowering of plants and

Assyrian design with a motif of pine cones c. 1000 B.C.

Winged deities worshipping a sacred tree – the Date Palm. Assyrian, c. 900 B.C.

the coming of fruit occasioned wonder and awe.

In primitive cultures trees have often been associated with religion, especially as holy trees and sacred groves. Sacred groves still exist in some countries where they often stand in isolation in an otherwise treeless and eroded landscape. They are usually associated with ancient burial places, often with one or more holy trees to which people offered cloth and articles of clothing to cure sickness or disease.

In ancient Egypt the Sycomore Fig was regarded as the dwelling place of Isis, Nut and Hathor. The goddesses disguised as the Sycomore Fig made tree and deity one. It is not without reason that this tree is associated with goddesses. Like other fig trees it contains latex and there are pictures from about 1500 B.C. which show a Sycomore Fig suckling a mortal Egyptian

Opposite: a sacred grove in North Africa with a thick growth of Aleppo Pine. In the background the surrounding landscape is hostile, treeless and bare.

An Ice Age painting of a bowman from southern Spain. The Stone Age could equally well be called the Tree Age. The hunter lived to a large extent in the forest and his weapons and tools were mainly of wood. The trees also provided the fuel for fire which was essential to primitive cultures.

The death prayer. The Date Palm, as were many plants and animals, was sacred in ancient Egypt. Tomb painting, c. 1400 B.C.

116

Mycenaean pottery displaying palm motif, from Crete, 1600–1200 B.C.

Above: Apollo with bay-leaf crown. Greek coin c. 360 B.C.

Opposite: Apollo with Daphne. Ivory relief carving. Byzantine c. A.D. *500.*

king to prepare and strengthen him for his work in life. In Greek mythology Zeus was associated with Oak, Apollo with the Bay Tree, Athene with the Olive and Aphrodite with Myrtle. The Dryads were wood nymphs who danced round their oak trees with oak leaves in their hair. The life of each nymph was associated with her own tree and ceased when the tree died.

According to the myth of Apollo and Daphne, as he pursued her she called upon the earth mother Gaea for help. Gaea made her disappear into the ground where a Bay Tree grew in her place. There-after Apollo regarded the Bay Tree as sacred.

As myths and legends were disseminated and handed down they were depicted in many ways – as direct illustrations of written or spoken words or as more or less abstract symbols as decoration. Perhaps the Adam and Eve motif is the best known as a text illustration to the Fall of Man in Genesis 3: 'And the eyes of them both were opened, and they knew that they were naked; and they sewed fig leaves together, and made themselves aprons'.

Dionysus, god of wine, surrounded by vines and clusters of grapes. Painting on a Greek vase of about 500 B.C.

A pine cone mounted on the point of a 'thyrsus' carried by a female votary of Dionysus, called a maenad (literally: 'mad woman') from about 400 B.C.

The goddess Athene, according to Greek
mythology, brought the olive to Athens.
She is often portrayed bedecked with
olive leaves. This is how the Renaissance
artist Sandro Botticelli painted her about
1480.

Right: vines incorporated in a design on
the capital of a column. Church in eastern
Spain, fifth century.

Right up to the present day many drawings, paintings and sculptures show Adam and Eve with fig leaves. The same consolidation is apparent in the myths of other cultures.

In the countries of the Mediterranean there is still impressive evidence of the use ancient cultures made of plants to illustrate religious texts and as decorative motifs. Plants are reproduced on columns and friezes, in mosaics, on vases and on coins. The Greeks and Romans incorporated plants in their architectural details. These were copied in the British Isles in the eighteenth and nineteenth centuries – particularly by the Adam brothers, who made extensive use of the *Acanthus* motif.

Opposite: Adam and Eve. The pomegranate and the fig leaf. Woodcut by H. Baldung c. 1500. The apple that Eve offers to Adam is a pomegranate, not the eating apple we know today and as it was known in the Middle Ages. The apple would not have existed at the postulated time of Adam and Eve. The pomegranate was of great ornamental significance up to and through the early Middle Ages. It is depicted sometimes quite naturally, sometimes stylized and almost unrecognizable. Below is a pomegranate motif worked in velvet c. 1500.

123

One of the legends associated with sacred trees is that of the giant Christopher, patron saint of travellers. He could cross turbulent rivers with ease, and on one occasion he carried a child across such a river. The child was so heavy that Christopher could barely cross the water. The child was Christ, who bore the sins of the world. In payment for his services Christopher's staff became a fruit-bearing tree – the Date Palm. The woodcut is German, c. 1400, and is late evidence of the spiritual importance of trees.

Further Reading

1. American School of Classical Studies at Athens, *Garden Lore of Ancient Athens*, Princeton, New Jersey, 1963.
2. Bramwell, D. and Z. I., *Wild Flowers of the Canary Islands*, Stanley Thornes (Publishers) Ltd, 1974.
3. Corner, E. J. H., *The Natural History of Palms*, Weidenfeld & Nicolson, 1966.
4. Feinbrun, N., and Koppel, R., *Wild Plants in the Land of Israel*, Hakibbutz Hameuchad, Israel, 1960.
5. Kunkel, G. and M. A., *Flora de Gran Canaria*, Vol. 1, Ediciones del Excmo. Cabildo Insular de Gran Canaria, Las Palmas, 1974.
6. Noailles, Le Vicomte de, and Lancaster, R., *Mediterranean Plants and Gardens,* Floraprint Ltd, 1977.
7. Polunin, O., and Everard, B., *Trees and Bushes of Europe*, Oxford University Press, 1976.
8. Polunin, O., and Huxley, A., *Flowers of the Mediterranean*, Chatto & Windus, fifth impression, 1974.
9. Polunin, O., and Smythies, B. E., *Flowers of South-West Europe: a Field Guide*, Oxford University Press, 1973.
10. Täckholm, V., *Students' Flora of Egypt*, Cairo University, second edition, 1974.
11. Walker, W., *All the Plants of the Bible,* Lutterworth Press, 1958.

For the wild flowers of the Mediterranean, 8 is recommended as the best introduction. More detailed treatments are given in 2 and 9, but of the works listed here only 10 will cover *all* the plants that grow wild in the country. Such technical and comprehensive works are known as Floras, and are listed in the bibliographies of 8 and 9, but may prove difficult to the beginner. For trees and bushes only, 7 is recommended.

Index

Page numbers in italics refer to pages with illustrations only

127